FEATS of CLAY
BRISTOL

OLIVER KENT

STEPHEN MORRIS

First published in 2025 by Stephen Morris
www.stephen-morris.co.uk

Text and photographs © Oliver Kent
additional photographs © Heather Kent and Stephen Morris

ISBN 9781068303937

British Library Cataloguing-in-Publication Data
A catalogue record for this book is available from the British Library

All rights reserved. Except for the purpose of review, no part
of this book may be reproduced, stored in a retrieval system,
or transmitted, in any form or by any means, electronic,
mechanical, photocopying, recording or otherwise,
without the prior permission of the publishers.

Cover: ss *Great Britain* 'Great Tile Mural'

Contents

5	The Ceramic City
13	Around the City
90	Some Bristol Potteries
92	Contemporary Potters, Brickworks, Postscript
94	The Cotham Hill Camel
95	Lost Ceramics – the one that got away
96	More to Discover
97	Acknowledgements
97	The author

A terracotta dragon from the John Street façade of the Everard Printworks.
Now in the foyer of the Clayton Hotel Bristol, 35-37 Broad Street, BS1 2EQ

The Ceramic City

In some cities bricks and tiles are a major influence on their appearance. New York and Chicago would not be themselves without their heritage of moulded bricks and coloured tiles; Lisbon is vibrant blue and white and more. The towns of England's south coast have their green and blue roofs and Edwardian tiled pubs. In Bristol you have to work harder, to explore, but it is worth it for the variety of treasures that are out there.

Despite a long history of pottery-making and a vast brick and tile industry in the nineteenth century, Bristol is by nature a stone city with ample supplies of building stone easily accessible. A walk across the city looking at older vernacular buildings is like a geology lesson; the choices of stone often reflect the underlying rocks. Brick only became common as the city expanded from the industrial revolution onwards and then mainly for worker-housing and industrial buildings. The geology is as complex for clay as it is for stone; bricks and tiles vary in character and colour. Look up and the chimney pots are often yellow but red and orange too.

Different clays behave in different ways depending on their chemistry and how they are fired. As a result, red and yellow surface clays from north and west Bristol become porous terracotta red. Grey clays from Bedminster and Kingswood associated with the coal-measures become hard buff and yellow. At higher temperatures the grey clays become impermeable and can be used to make firebricks and furnaces. Fired hotter the red clays become purple and black and impervious to water. Hotter again and they start to liquify – Avon mud turns to a rich brown-black glass. Potters and brickmakers know how to exploit these things.

In many parts of the city old clay pits lie forgotten. Going out of Bristol on the A370, as you clear the city boundary, the trading estate on the left is not lower than the road by accident but because it was a huge clay pit for brick

Potters and brickmakers have left their mark across the city. These bumps and dents at Clay Pit Road, Durdham Down are back-filled pits, probably for the Sugar House Pottery in Westbury-on-Trym. Owned by poet Samuel Taylor Coleridge's father-in-law, Sugar House specialised in sugar-moulds and chimney pots

production in the nineteenth century. On the northern edge of Durdham Down, Clay Pit Road is a clue. Clay digging has left the ground lower than the roads and there are hollows in the surface – the result of the backfill of small pits settling over the last 200 years.

As a result, teasing out its ceramic gems becomes one of exploring and constantly being on the lookout. Bristol's main Edwardian brewery, Georges, went for stone and red brick rather than the glazed tiles fashionable in some other areas of the country. A gorgeous green-tiled pub like The Hare on the Hill in Kingsdown is a rare trophy.

Rather than be disappointing, it makes the hunt more fun. You start to notice the local brickmakers' enthusiasm for trying to outdo the stonemasons. Moulded fireclay cornices and brackets and columns and all the details that a builder might want in stone were made available by brickworks in Kingswood and Bedminster. In the later 1800s many small terrace houses were built with their window and door surrounds set out using the simpler mouldings but every now and then you find a builder going the whole hog as on the curved terrace of shops on the Clifton side at the top of Blackboy Hill. Modelled heads, coats of arms and animals sometimes turn up in unlikely places like the shop on the end of Raleigh Road in Southville. In Vale Street, Totterdown (famously the steepest street in England) the quirky assortment of tiles on the top few houses look as if they might have come from the builder's leftovers box.

Some large nineteenth-century public buildings were made entirely of brick and terracotta but Bristol still tended to prefer stone. The former Prudential Insurance Company offices in Clare Street or the WD & HO Wills factories in Bedminster are prominent examples. The highly finished terracotta ages better than the stone so many of the buildings using these materials look surprisingly fresh for their age. The range of possible brick colours allowed architects and builders to have fun too. The use of iridescent black brass manufacturing slag cast into blocks as a building material is a purely local phenomenon.

The use of glazed bricks became popular in the late nineteenth century not least for their ease of cleaning for specialist businesses like butchers and fishmongers, and for public lavatories. The potential for colour and decoration was irresistible and thus brightly coloured ceramic shops and pubs appeared. Designers took it further: WJ Neatby's 1901 façade for the Everard Printworks in Broad Street is a spectacular example of Doulton at their most exotic. Reimagined for the 1930s, smooth tiles provided a sleek surface for the likes of the Art Deco Odeon Cinema in Broadmead.

The decline of the smaller hand-made brick producers through the twentieth century and the dominance of a small number of big companies using a limited number of clay sources has had a dulling effect. The bland buff and grey tones of suburban housing estates in places like Bradley Stoke are typical of the resulting standardisation, no longer reflecting local materials but looking the same as estates across the country. Nonetheless, since the 1990s there has been a resurgence of interest in the potential of bricks and tiles. Handmade bricks have become a niche product and traditional bricklaying has become expensive, but there is an increasing interest in moulded elements and the creative use of varied colours.

The costs associated with on-site bricklaying have led to brickwork being laid off-site, cast into concrete blocks or attached to metal frames so that they can be craned into position and bolted on. Asda in East Street, Bedminster was put together this way. By using red and cream brick and features like arches and window forms that echo its neighbours the former Wills Tobacco Factory No.1 and the Bedminster Public Library, the architects sought to blend it into its location.

The later twentieth century has seen a growth in the use of brick as a material for public art, often in the form of reliefs on buildings, picking up where architects left off in the early part of the century as modernism moved away from such things. Moulded and handmade pieces have been joined by brickwork carved with power-tools. Using the bricks themselves unaltered is less common but the giant fish at the former Fishponds railway station stands proud and the shadowy night-fighter on the side of Blenheim Court on Filton Avenue is a subtle, skilful piece of bricklaying commemorating the close relationship of Filton's people with the wartime aeroplane industry. Most recently, street artists have discovered ceramics. Bristol's Upfest Graffiti Festival is a major showcase for new street art and tiles, ceramic mosaics and even fully three-dimensional pieces have been appearing regularly particularly south of the river. They are even ceramic tags if you look for them.

Urban Foxes. Chinagirl Tile for the Upfest Graffiti Festival, 2016,
Hen and Chicken, 210 North Street, Southville BS3 1JF

It is not all about buildings and architecture though. Sometimes the clay itself is the feature story. Soft bricks invite scratched graffiti – a brick wall in Henbury Road, Westbury-on-Trym records the presence of American servicemen during the build-up to D-Day in 1944. The Avon is one long line of clayey mud running right through the city. The artist Richard Long has made it one of his primary materials for recording his movement through the landscape. During Lockdown, many artist potters found, as we all did, that their normal working practices were disrupted or ground to a halt. There were even clay shortages. Many potters and students took to sourcing local materials and searching out local clay supplies. Gardens, fields, builders' skips and riverbanks were combed for 'wild clay.' They were rediscovering the variety of our geology in the process.

Bristol is no brick city but it has colourful and varied ceramic treasures to seek out. The following are some of my favourites – go and find yours.

Flyposted ceramic mosaic. Anonymous. c.2015.
Leonard Lane Bristol BS1 1EA

National Cycle Network

Outside the entrance to We The Curious, Bristol's science museum in Millennium Square, a huge relief tiled map celebrates the National Cycle Network.

Bristol is the home of Sustrans the charity that has promoted cycling and sustainable transport since the 1970s. In the early 1980s they successfully established the Bristol to Bath Cyclepath and on the back of that campaigned successfully for a national network of designated cycle routes. With substantial government support the network was launched in 2000 and has become a hugely valued asset.

To celebrate the opening of the network, three large tiled maps were commissioned from Bristol artists Marion Tucker, Sue Ford and Carol Arnold. One was sited at the hub of the network in Centenary Square, Birmingham in 2000, one in Belfast in 2001 and this one in Bristol, the home of the project, in 2003.

The tiles are made of stoneware paper clay. In the 1980s some potters began mixing paper pulp with clay to create a strong clay-like material that had next to no shrinkage and could be joined even when almost dry. The downside was that the pulp rotted if it was left wet for any length of time and created a stinking toxic mess.

We The Curious
Millennium Square,
BS1 5DB

Edwardian Butchers' Shops

Collard's butcher's shop in North Street, Bedminster, Bannister's in Worrall Road in Clifton and Woolston's in Lower Redland Road are classic examples of the Edwardian butcher's shop. Huge plate glass sash-windows opened to reveal marble slabs and displays of meat open to the street. The interiors were covered with white (or at Collard's decorated) tiles to emphasise cleanliness, together with sawdust on the floor. Also common is the use of green glazed bricks below the marble and on any flanking surfaces. Woolston's shop in Lower Redland Road is a particularly rich sea of glassy green speckled by the coarse fireclay of the bricks themselves. Bannister's bricks have the name very precisely written across them. The letters are drawn using tube-lining, a process similar to cake icing using a small squeezable bottle of colour. It was often used on floral fireplace tiles and is a key feature of 1930s' art pottery by the likes of Moorcroft and Charlotte Rhead. Drawing such precise serif lettering would require a very steady hand – not for the faint-hearted.

Collard's also features Aldred Daw Collard's name but in a wonderful hand-drawn style. He had rebelled against his father and set up independently of the family firm so his identity was particularly important. The letters are free-hand drawn using underglaze colours (prepared colours applied under a clear glaze). The letters are in a fairground style, naively drawn to imitate green enamelled brass relief. The brass edges have gone wildly wrong at times – C and R are particularly Escher-like. One of the Collards had a second string as a poet, specialising in meat-related verse!

R Bannister	A D Collard	Woolston's
(now the Emzi Café)	(now Friendly Records Bar)	(now The Bristol Artisan)
3 Worrall Road	57 North Street	3 Lower Redland Road
Clifton	Bedminster	Redland
BS8 2UF	BS3 1ES	BS6 6TB

Temple Meads Station Extension
PE Culverhouse, 1935

When glamorous maroon and cream vintage trains pull into Temple Meads Station, they are greeted not by the Victorian splendour of the great engine shed that reaches across platforms 2-5 but by shiny purple Art Deco tiles announcing their arrival in Bristol on the glossy glazed ceramic waiting and refreshment rooms of the open platforms to the east. Quite Agatha Christie, Miss Marple and cream teas.

Built in several stages starting with Brunel's original terminus in 1840, by the 1920s the station was much too small for a large city like Bristol. In 1932 a major expansion project doubled the number of platforms and provided an underpass linking them all to the ticket hall. The increase in lines required the widening of the bridges over the Avon and the New Cut.

Around the same time Great Western Railway was busy building new Art Deco stations for Cardiff and Leamington Spa, and office blocks at Paddington. But dominated by Portland Stone, none of them have the pizazz of Temple Meads' cream glazed ceramic walls and panelled waiting rooms.

The raised lettering on each platform is glazed a rich purple, also used for tiled poster frames facing the carriages. It is a nice touch that the railway has recently commissioned artists to design posters of local views to fill them.

Bristol Temple Meads
BS1 6QF

The Hare on the Hill

The Hare on the Hill in Kingsdown is one of the few pubs in Bristol to sport a brightly coloured tiled façade. Many breweries used them in the early part of the twentieth century as a marketing tool and in combination with similarly decked out interiors to impart an air of cleanliness.

Until recently known as The Mason's Arms, the pub clearly had a makeover soon after 1900 and went for Art Nouveau tiles in a big way for the single storey extension to the front of the old cottage. Sitting on a ledge on the steep hill the tiles vanish into the pavement at one end whilst sitting on a raised plinth at the other. It looks for all the world like a brightly painted Lisbon funicular car, especially at night with all the lights on. All the more welcoming for the bright paintwork and the cheerful mural on the Cow Byre next door.

Elsewhere, tiled pubs often promoted brewery names and beer brands in large letters and the more exotic ones had elaborately painted tiled signs. Whoever extended the front of the Mason's Arms went for Art Nouveau in green, yellow and cream but the lack of brewery signage suggests an independent owner (compare it with The Punch Bowl in Old Market). The decorated tiles are moulded, not the more expensive hand-decorated ones, although the coloured glazes would have been applied by hand. The main turquoise green glaze is sumptuous. The beer's good too.

The Hare on the Hill
41 Thomas Street North
Kingsdown
BS2 8LX

The SS Great Britain 'Great Tile Mural'
Pupils of Ashley Down Junior School; Bishopsworth Junior School, Teyfant Community School, Holy Cross Primary School and Hillcrest Primary School, 2004

Children and clay go together like cats and ping-pong balls. In 2004 children from five primary schools across the city came together to celebrate the SS *Great Britain* by making a large tiled panel. One hundred and twenty relief tiles are linked both by the colour scheme and the scale of their designs. The children had evidently spent time looking very closely at ironwork and decorative details with an emphasis on cogs and levers and chains.

The SS *Great Britain* was rescued and brought back to Bristol not just because it was an old Bristol-built ship but because it was an important part of the history of maritime engineering and one of Brunel's major achievements. The children's panel works as an artwork but it also works as an idea, celebrating the ship not for its own sake but for its engineering and reminding us that this is quite literally the place where local people physically did that work. Archaeology preceding the construction of the flats on the other side of the road revealed the workshops and the bases for the machinery they used.

It is good to see that the SS *Great Britain* Trust saw to it that the panel was retained when the building was renovated in 2014. So often community-based projects like this are seen as temporary and disappear quietly.

1 Brunel Square
BS1 6JR

Samuel Shield's (former) Filton Laundry
Washing Machines, Rodney Harris, 2006-8
Washing Line, Rodney Harris and Valda Jackson, 2013

The Art Deco offices of the Bristol Aeroplane Company stand proud at the top of Gloucester Road North looking out over the factories and the airfield. Pegasus bursts from the first floor and at rooftop height a life-size aeroplane forms a frieze. This is a celebration of the industry but it also reflects a powerful local pride. Across the road the Filton Laundry was also an important local employer. Later, as part of Filton Technical College, its large open factory spaces housed trainee gas-fitters and art students who remember it fondly. In the early 2000s the site was cleared and redeveloped as a shopping centre. Artist Rodney Harris was asked to design a sculptural celebration of the past history of the place and wittily brought together the two aspects. Harris and his colleague Valda Jackson take raw bricks from the production line, carve and model onto them to create reliefs that can then be fired with the standard bricks and built into the walls of buildings.

Here the first bay of the wall of the passage between the retail units on the Gloucester Road North side has been reworked as a row of five life-size washing machines as if it was occupied by a launderette. The machines are glazed shiny bright white and are quite disarming especially from the road. We often think of clay as an artist's material, only suitable for small objects that might hold water. These pieces emphasis that we can make ceramic objects as big as we like using multiples and still at least metaphorically hold water.

When the centre was extended in 2013, the developers commissioned a second panel. On the north wall, Valda Jackson created a line of washing blowing in the wind, growing subtly out of the yellow brick wall overlooking the car park.

Shield Retail Centre
Link Road
Filton
BS34 7BR

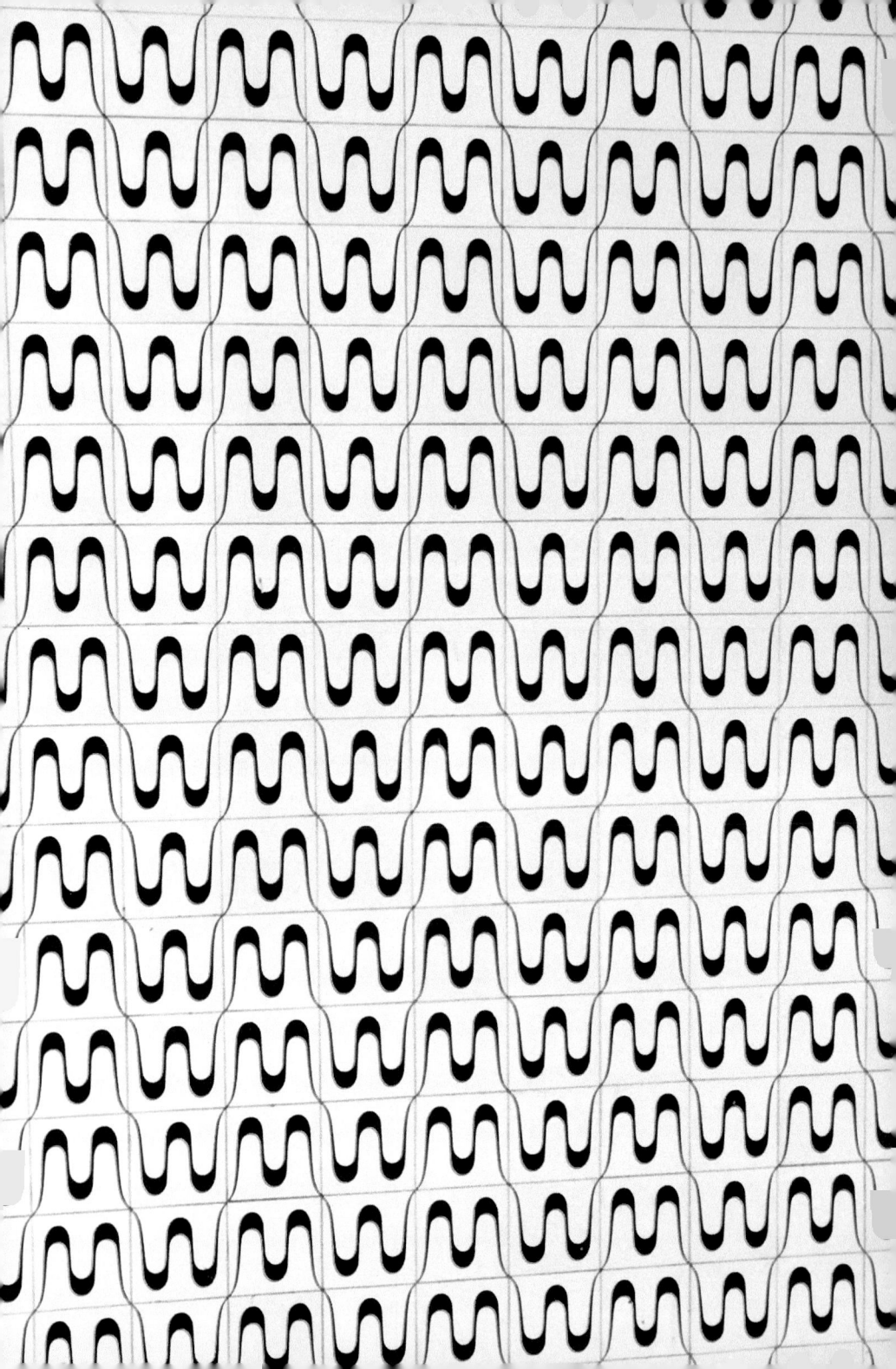

Brandon House
City Architects for Bristol County Borough Council, 1959

The entrance to the flats of Brandon House is flanked by an elegant black and white tiled screen wall. The design is probably by Peggy Angus for Carter's of Poole. Poole tiles were often used by Modernist architects at this time, but they are rare in Bristol.

The 50s saw a resurgence of mural design and tiling, with contemporary tiles on public housing, public buildings and shopping precincts. The Transport and General Workers' Union building in Belfast has a dramatic wall depicting workers and industries up the full height of the building. Bristol was more cautious and these tiles are not so assertive but, even so, there is power in their simplicity.

At the foot of Jacob's Wells Road, the twin blocks of St Peter's House also date from 1959. Between the two blocks is a more adventurous, curved panel of tiles which, like the TGWU mural in Belfast, rises from the first floor to the top of the building. Sadly, the tiles have long since been obscured by paint to leave a giant white rectangle. St Peter's House is easy to ignore, but walk through the passage under the left-hand block and onto White Hart Steps to see Bristol's most exciting brutalist fire escapes – a wonderful, hidden treasure.

Brandon House
Jacob's Wells Road
Hotwells
BS8 1DT

Dame Emily's Frog
Jo Young, 2004

Frogs are a thing in children's play areas, generally fibreglass ones mounted on springs with a certain bouncy logic. Not here. The frog in Dame Emily Park in Southville is not that kind of frog at all. It is a giant, of handmade stoneware with a rich waxy, froggy green glaze and bright eyes. Known as Granny Frog, she sits alert beside the play area overseeing the children's goings on. Big enough to sit on but tough enough to survive it, she may be just about to hop away.

The park was laid out over the pithead of the Dean Lane Colliery after it closed in 1906. Below ground the colliery was huge, extending as far as Temple Meads Station. By 2000 the it was well-known for The Deaner, a concrete skatepark, but the rest was neglected.

Each generation has its own view of what a park might be. For the mining community of 1906 a bandstand was a great asset. The octagonal platform in the skatepark is the imprint of the bandstand that sat on top of the capped mineshaft as a memorial to those miners who lost their lives here. By the 1980s the park had become a focus for the skate-boarding community. Our frog was commissioned in 2004 by the Dame Emily Park Project as local parents strove to revitalise a valued but neglected resource. Now, Granny Frog presides over a green space, a garden and a playground.

Dame Emily Park
Dean Lane
Southville, BS3 1BS

The Black Castle
James Bridges, *c.*1765 for William Reeve

When clay is fired high enough it melts — there is no precise boundary between ceramic and glass. The Black Castle glistens like a bluebottle's back in the sun, its shiny blocks contrasting against the pale Bath stone detailing. It was built for William Reeve in the 1760s as the stable block to his new house, Arnos Court. The blocks were cast from the sticky liquid slag from the copper smelting works on the north side of the Avon. The colour is largely iron mixed in with silicates and other unwanted components of the copper ore. There is very little copper in it, although if you search you will find some of the blocks leaching green and yellow stains into the white mortar. Friday afternoon smelting, perhaps! In effect the slag sits somewhere between a red clay fired to the point of melting and an iridescent black glaze.

Eighteenth-century industrialists and landowners were very proud of their new technologies and unashamedly celebrated them. Above the copper works on the north bank at Warmley, owner William Champion built his house to overlook the works and used the mill pools and races that powered them to enhance his gardens. The use of the slag was not only a dramatic aesthetic statement but one of industrial success and the drama of progress. Today as a pub, the Black Castle still stands proud, defying the grizzly retail-park in which it now finds itself.

St. Philips Causeway
Arnos Vale
BS4 3BD

A Loire Château in the city: Prudential Assurance
Alfred Waterhouse, 1899

The Prudential offices stand out amongst the commercial buildings in the centre of Bristol, a shiny rich red terracotta French château amongst the muted colours of its neighbours. St Stephen's church next door is separated from it by the churchyard and they sit surprisingly well together. However bold, most commercial offices in city centres are effectively façades competing in a row of others. Here the narrow medieval block-end site provided the architect Alfred Waterhouse with the opportunity to work fully in three dimensions. The great pyramidical slate roof and the round corner turrets are vital in making it as much fun as it is. It peeks enticingly between the grey offices on St Augustine's Parade drawing you in to the little oasis of the churchyard of St Stephen's.

The building is faced with moulded and polished terracotta blocks. These were made in plaster moulds by hand and the larger ones are hollow, the finger prints of their makers concealed inside them. If you look closely in the corners of window mouldings and details you will find traces of modelling tools used to tidy up the castings. Every inch of the clay surface has been lovingly made by real people.

Waterhouse designed offices for the Prudential across the country but the Bristol site allowed him the most freedom. He is best known for a much larger and more prestigious terracotta building, the Natural History Museum in South Kensington. It was made the same way, inside and out including all the decorative detail and those monkeys that run up and down the columns inside. Apparently, the firm that contracted to supply the ceramics went bankrupt having agreed to a fixed price for the job.

17-19 Clare Street
BS1 1XA

The Café Wall Illusion

In the early years of the 20th century 19-20 Perry Road, the corner shop at the junction with St Michael's Hill, was given a full makeover as a butcher's shop with a shiny fashionable tiled façade. Unusually the tiles were set out in stretcher bond like a single-brick wall or Lego, each tile spanning two below it equally. The alternate green and cream rectangles created a strong vertical beanstalk pattern. It looked good and it is not clear whether the resulting odd visual effect was immediately obvious.

By the 1960s the butcher's shop had closed and the building was remodelled as a café. Although the interior of the building changed, the tiled front remained intact and it was probably as a result of repointing with a dark coloured grout that things began to get strange. The horizontals steadfastly refuse to be parallel even though you know they must be. One row dips to the left, the next to the right and so on. Try drawing the pattern out on a piece of graph paper – the same thing happens.

It was in 1979 that Professor Richard Gregory of the Brain and Perception Laboratory at Bristol University christened it the 'Café Wall Illusion.' After a member of the lab pointed it out to him, he set out to understand why our brains do not seem to be able to cope. The conclusion is essentially that the strong contrast between the colours and the relationship with the dark line of grout leave our brains unable to work out exactly where the edges are. The brain tries but comes up with contradictory results. Well I think that is it.

Not content with being well known for illusions, in the 1980s as Special K café it was the hangout of The Wild Bunch who later became Massive Attack.

808 Café Bar
19-20 Perry Road
BS1 5BG

Potsherds Between the Cobbles

The wide expanse of quayside beside Redcliffe Bridge has been awaiting development for years. The ruined buildings at the back are the last vestiges of a busy industrial area making pottery and glass. The pottery industry in Bristol took off in the late seventeenth century. From around 1700 the Frank family's potteries extended from their house in Redcliff Hill to the wharf below, specialising in blue and white delftwares for the table and the kitchen. They also made salt-glazed stoneware tavern mugs and yellow slipware cups and dishes.

Despite attempts to develop porcelain, Bristol potters were not able to capitalise on their early success. Competition from Chinese and English porcelains and creamwares, coupled with the rise of Stoke-on-Trent, led to a steady decline as the century progressed.

The Franks relocated to Temple Back in 1777 and the Redcliff buildings stood empty before being used for a time as a prison for sailors captured during the American War of Independence. In 1779 prison reformer John Howard found 150 French prisoners incarcerated in the pottery buildings.

The wharf's history lurks under the cobbles and tarmac, peeking out when the surface gets disturbed. Little bits of stoneware tankard and pieces of unglazed 'biscuit' delftware appear along with the odd blue and white delft tile sherd or fragment of plate. The nicest pieces are the glassy fragments of salt-glazed saggars – pottery containers used to protect the stonewares in the kiln. The potters must have dumped a lot of their own rubbish on site or else provided it as hardcore during quayside construction work.

Redcliff Wharf
Redcliffe Way
BS1 6SR

The Mission House and Reading Room
c.1870 for Philip Henry Vaughan

The Mission House and Reading Room is a bright surprise hidden away off Blackboy Hill near Quarry Steps. It is not in the architecture books but it is a little cousin to the magnificent Granary on Welsh Back – Bristol Byzantine at its richest, playing with red, cream and black brick and the skills of the bricklayer. Blackboy Hill was once dominated by quarries and at least one huge hole survives containing the Spire Hospital, others form cliffs between Upper Belgrave Road and Worrall Road. There are smaller ones along the hillside as far as Redland Green. This is the home of the Bristol dinosaur, *Thecodontosaurus*.

The building must have been a dramatic addition to what was then a hamlet of small stone houses (and at least six pubs). It was built around 1870 for the parish church of St John to provide religious, social and educational opportunities to the local community. The founder was Philip Henry Vaughan of Georges Brewery. The Granary dates from 1869 and was designed by Archibald Ponton for the grain merchant WK Wait. The Waits also had a connection to St John's and were perhaps involved in some way too.

The great joy here is the playful use of coloured brick. The front is dominated by an ornate arcade of floor-to-ceiling windows on the ground floor with their patterned arches separated by circular lights. The building runs right through the narrow block to a steep back lane. Here where you might expect utilitarian plainness, the fun continues. At the back of the Reading Room the black and cream arches of five gothic windows intersect like a Fair Isle sweater pattern below a huge round window. The shaping and cutting of bricks to make it all work must have been a labour of love and all done beautifully. Many of the motifs on the front can be found on the Granary but the back of the Mission House is all its own: joyously asymmetric, whereas the Granary is maths and neatness.

Now flats and nursery
Richmond Dale
Clifton
BS8 2UB

Rodway Cloister
Courtauld Technical Services, 1966

A frequent feature of 1960's Brutalist buildings is the use of concrete, brick and sometimes glass to form screens and dividing walls. The aim is often to allow light into confined spaces particularly at the base of buildings. The Dove Street flats in Kingsdown use perforated concrete blocks to allow light and air into the area around the street level access to the lifts.

The University of Bristol's Chemistry Building is unusual amongst the main campus buildings on St Michael's Hill in that it incorporates a generous amount of public space on two levels. Built in the early 1960s, the three main blocks surround a large plaza with an atrium-style cloister underneath it. Buried in the steep hill-slope, the cloister could be claustrophobic, relieved by a tree and daylight breaking in through the large central lightwell from the plaza.

At the rear of the cloister where it is cut into the slope, the back wall is concealed by a screen made of shaped black engineering bricks leaving a deep space behind it. The effect is to soften the space and to render its real size indeterminate. The plaza above feels bleak despite the spectacular concrete mural and the expanse of sky. Down here is a Brutalist black, white and grey space of calm.

University of Bristol,
School of Chemistry
Cantock's Close
BS8 1TS

West Country Ales
Royal Doulton, c.1960

In today's pubs, clubs and bars it is pretty much the rule that every drink is served in a suitably branded glass. In the 60s and 70s the barman was more concerned as to whether you wanted your pint in a mug or a sleever. The marketing novelties were much more fun and the potters were the go-to people. Carlton Ware in Stoke-on-Trent made toucan table lamps for Guinness to create atmosphere at the bar. Soft drinks brands provided jugs and every drink and cigarette manufacturer had its chunky ashtrays. Pountney's Bristol Pottery in Fishponds was a major supplier.

The long association of pubs and ceramic tiles, inside and out, is partly about cleanliness. It perhaps also resulted in a familiar relationship between the breweries and the potteries. When you wanted jugs and ashtrays and such like, you knew who to go to. Post-war, establishing your brand on a pub was unlikely to involve a full terracotta or glazed-tile remodelling and new buildings were distinctly understated in that regard. It seems one solution was to use the familiar ceramic tile, but as a label.

West Country Ales were a Gloucestershire brewery based in Cheltenham. The Doulton stoneware sign dates from the late 50s or early 60s just before they disappeared into Whitbread's. At the Palace Hotel it has been cut into the rusticated stone of the Victorian façade – their pub not Courage's who dominated Bristol. On 60s' new-builds these tiles were carefully set in the brickwork to give an air of permanence and heritage.

Palace Hotel
West Street and Old Market
BS2 0DF

Edwardian House Tiles

Suburban Bristol at the turn of the twentieth century was largely made up of streets of terraced and semi-detached stone-faced houses with Bath Stone bays, windows and door surrounds. To give them a bit of additional individuality the builders sometimes used the rectangular panels below the windows for decoration. At the simplest the panels are filled in with red brick or some carved Bath stone but around 1890 with the growing popularity of polished terracotta, the builders seem to have seen an opportunity. Inspired by the Arts and Crafts Movement, panels of chunky red flowers began to appear.

Most of this was pretty discreet but now and again they took it a bit further. In Northumberland Road in Redland the uprights of the bay windows have full height sunflowers growing up from urns inset into them, while rows of shorter sunflowers and lilies occupy the horizontal spaces below the windows.

Sometimes the results are more eccentric; in Vale Street, Totterdown segments of circles suggest an end of line batch of tiles belonging to a much larger pattern. In Linden Road in Westbury Park, a small group of houses are decorated with blue floral tiles of the kind more usually associated with fireplaces – not intended for outdoor use, some are slowly being destroyed by winter frosts.

High Kingsdown
Whicheloe MacFarlane and JT Group, 1970-74

The estate between St Michael's Hill and Alfred Place won awards in the 70s for its innovative approach to urban housing, maintaining density without relying on high-rise and seeking to create community, something high-rise had failed to do. The model was influenced by earlier designs by architects like Eric Lyons of Span, who sought to isolate homes from traffic, incorporate green spaces and bring people together. The ultimate source was the Bauhaus. The zig-zag layout of the estate allows the houses to avoid overlooking one another and allows the light in. High garden walls provide further privacy.

The design does not allow for individuality though, a standard pattern of black paint, white cladding and golden yellow brick is broken only by planting. The London brick on the other hand is wonderfully colourful, each one mottled and striped by the vagaries of the firing process and the impurities in the clay. It is curious though, in a city where brickwork colour has been driven by local materials in a range of reds and creams, to suddenly move to an orange-yellow London brick.

From the 1950s onwards, brick buildings have become increasingly uniform in colour across the country, pale browns and buffs predominating. Bradley Stoke is typical. A significant cause of this is the replacement of small local brickworks with a few large brick manufacturers run by large national companies in the years after the second world war, a reduction in the variety of clays used combined with a move to national distribution networks. A small national palette replaced an idiosyncratic localised one. Latterly a renewed interest in brick is slowly changing things.

Bounded by St Michael's Hill,
Myrtle Road and Alfred Place
Kingsdown BS2

Salt and Hygiene

Some of the east Bristol brickworks using the coal-measure fireclays specialised in salt-glaze, primarily for drainpipes. The use of salt thrown into the kiln to provide an instant glaze was discovered in north-west Europe in the Middle Ages. The result was an incredibly tough ceramic that was easy to keep clean and resisted acids and unpleasantness. Thick acrid fumes from the kilns were a distinct downside and by the mid 19th-century most salt-glaze potteries were exiled to the edges of towns. An act of parliament controlling them was partially triggered by the fumes blowing across the Thames and into the Houses of Parliament from Doulton and Co's stoneware pottery on the opposite bank in Lambeth.

The upside was the handsome nature of the glaze itself, especially the rich browns of clays with even a small amount of iron in them. There are some beautiful drainpipes and U-bends to be found! Salt-glazed bricks were often used inside and outside buildings like factories where dirt was seen as an issue. Combined with white glazed bricks, they are often found in institutional and public lavatories.

The house on the end of Chessell Street in Bedminster, built almost entirely of iridescent salt-glazed bricks, is a shiny eccentricity. Whether there was some noxious neighbour on Luckwell Road or whether it was an aesthetic decision is anyone's guess. Even the gateposts are made of black engineering brick to emphasis the resistant nature of the property.

Luckwell Road
Bedminster
BS3 3EW

Swift's Cold Store

Glazed brick and tile façade for Swift and Company Ltd, *c*.1910.
Tiles probably by Carter & Co of Poole, Dorset

It is easy to miss this building, set back from the road in a back street at the far end of Old Market. A rare green-glazed façade for an Edwardian frozen meat warehouse. In its current state, embellished with graffiti it looks a little unloved but glazed ceramic isn't easily damaged by paint.

Green, glazed brick columns divide the delivery bays and support a frieze of faience tiles with the company name in large yellow letters and above a red, white and blue 'S' logotype. Swift and Company were a very large American meat processing company based in Chicago; their success driven by bulk refrigerated distribution. Their British subsidiary opened in 1891 in Smithfield, and by 1914 they claimed to have wholesale branches in all the principal towns. Supplied by huge abattoirs in the US and Argentina they undercut small family butchers and helped to establish branded chains of shops. Keen to make an impression, at some point they commissioned an aesthetic overhaul of the branches; the façade appears to have been stitched onto an existing building.

Truck delivery bays and big logos, this is an attempt to look modern, American and different – totally at odds with the white interiors and pictures of cattle and art nouveau swags of flowers so common on high street family butcher's shops like Collard's in Bedminster. This might look 'Art Deco' and romantic now but that was not at all the intention at the time. There was a similar building in Bournemouth and it is very likely the tiles were made by Carter & Co of Poole, who were well known for this kind of work.

Bragg's Lane
St Judes
BS2 0DN

A Terracotta Welcome

Shopfronts with recessed entrances always make you welcome in some way – terrazzo, mosaic or encaustic tiled doormats are wonderfully varied, sometimes including the name of the original shopkeeper. Some are hearthrug-like and friendly, others downright weird. One of the oddest is the glass-block cellar-light with a ceramic tiled border that greets customers to the British Heart Foundation bookshop on Blackboy Hill. It's particularly disconcerting when the basement lights are on. What Victorian lady customers thought of stepping across an illuminated glass panel is unrecorded!

The ceramic tiles used are the same ones found in the foyers of Victorian public buildings and the hallways of Victorian semis. A range of unglazed triangles, rectangles and squares made from specially prepared coloured clays were designed to fit into almost any space, the patterns embellished with fancier tiles with inlaid designs. These were usually simple versions of gothic patterns derived ultimately from medieval tiles. All the tiles were dust-pressed using finely ground leather-hard clay rammed into moulds by a mechanical press. Hard-fired they are very robust and the better quality ones often look as good as new. Larger versions were popular for garden paths and have been making a comeback more recently.

There are thousands of variations: just look where you are putting your feet.

13 St Stephen's Street

This bright Victorian glazed-brick office building has a surprise: busts of Milton, Tennyson and Shakespeare prominent between the ground floor windows. Built in 1878, it predated the better-known ceramic façade of the Everard printworks in Broad Street by more than twenty years. The poets are a mystery, perhaps they reflect the tenants the developer aspired to attract.

Number 13 is covered front and back in yellow, blue, orange and brown glazed brick with distinctive matt terracotta window mouldings and modelled detail. The shiny façade reflects light into the dark street and lane and is divided by stripes of coloured brick and two bands of terracotta flowers. The unusual delicate orange is a lustred white brick. Weathering is gradually removing the colour; it must have been striking when fresh. The whole thing seems less like architecture, more 'look-at-me, come and set up business here'. By 1902 it was inhabited by solicitors, a stationer and an electrical engineer: a new science at the time.

A curiosity: St Stephen's Street is narrow medieval street running along the outer face of the city walls. This is the original line of the River Frome before it was re-routed along what is now The Centre. Behind the north side, Leonard Lane follows it, the alley behind the walls. A flight of steps to the right of number 13 leads up to the lane, a good six metres above the road. The buildings are barely one room deep – effectively the thickness of the medieval wall.

13 St Stephens Street
BS1 1EL

Ceramic Histories

The University of the West of England's School of Art and Design at Bower Ashton was built in the 1960s to bring together the scattered parts of the West of England College of Art that had outgrown its Victorian premises on Queen's Road. On the edge of Ashton Court, the building provided spaces for new courses including ceramics, graphic design and construction. Ceramics was led by George Rainer and set out to be fresh and contemporary with an emphasis on clay as an artist's material. Despite its successes the department closed in the early 2000s and during alterations George's old office window was blocked up. Rather than replicate the dull grey 1960s concrete bricks someone decided to use the salt-glaze kiln round behind the Construction Building to make their own.

The bearded face is modelled on a seventeenth-century German salt-glazed wine jug known as a bellarmine or bartmann. Much loved, they are said to be intended as a caricature of Cardinal Bellarmine, persecutor of protestants in north-west Europe in the late sixteenth and early seventeenth centuries. The cardinal would have been a child when the first ones were made but the nickname stuck in his lifetime.

More usually called bartmann (bearded man) jugs nowadays, this panel references the history of ceramics and uses it to caricature the stereotypical potter, wild and hairy, no doubt wearing a smock and covered in clay. As such it is a memorial to a successful course that fell by the wayside, although a small pottery now operates in the brand-new building opposite.

> University of the West of England
> School of Art and Design,
> Kennel Lodge Road
> BS3 2JT

Bristol United Press
Group Architects DRG, 1972

Hot chocolate brown engineering brick and plate glass made this building stand out dramatically, helped by the huge digital clock on its corner facing oncoming traffic in the underpass below. Digital clocks were all the rage in the 70s. Built as two massive blocks nestling together the left hand one, the Print Hall has now been demolished. The architects designed for this eventuality knowing that print technology would change and it would become redundant over time. The other reason for separating it was apparently the noise and vibration of the machinery: massive concrete foundations were provided to anchor them securely.

The shiny engineering brick seems both severe and warm at the same time – there is definitely something chocolatey about it. The windowless Print Hall was the most prominent aspect, the glass atrium and office tucked behind and shielded by trees. Bricks can soften a building in a way that concrete never can and the curved outer wall of the Print Hall and the rounded corners of all the edges of the building add to that softening effect.

Although the building is less prominent now, you can get a strong sense of its former drama when you get close to it and admire the tall round cornered stair towers. On the Broad Plain end a quirky, implausible brick 'fence' raised on little concrete legs sits on top of the wall screening the car park. The sections are actually tiled but the sense that rows of bricks are defying gravity remains. The building is a bit too playful to be called Brutalist.

1 Temple Way
Old Market
BS2 0BY

Broad Quay House
Aspects of Bristol History, Philippa Threlfall and Kennedy Collings, 1981

Broad Quay House sits on the end of the Centre at the top of St Augustine's Reach. Its waterside face is bland – in 1981 developers still assumed that the quayside was the back rather than the front of their buildings. But above the ground-floor windows on the land-ward side, fifteen ceramic tableaux liven up the façade and celebrate Bristol's industrial history.

The panels are by Philippa Threlfall and Kennedy Collings and are made up of cast and modelled earthenware tiles. Each has an individual medal-like roundel depicting a key trade or location, set between two heraldic sea creatures. Fourth from the right, 'Success to the Glass Manufactory,' celebrates an industry the city remembers proudly, portraying the iconic cone-shaped kiln once so familiar to Bristolians.

Many of the industries that made Bristol were smoky and smelly, a dystopia of chimneys and furnaces. In historic views of St Mary Redcliffe, the background is dominated by the great smoking pyramid of the nearby glass-cone – one of many. Elsewhere in Redcliffe the passer-by would encounter the bottle-shaped hovels of the nearby potteries. The cones and hovels sheltered the kilns and furnaces inside, and acted as chimneys by creating a draw. Their thin-walled tapered brickwork was the work of highly skilled men from specialist firms supporting the ceramic industries. Today, the base of the glass-cone behind St Mary Redcliffe is the only survivor and has been a restaurant for many years.

Broad Quay
BS1 4DJ and
Kiln Restaurant
Prewett Street
BS1 6PB

The Christmas Street Eye
Beau Stanton 2017

This is an intriguing piece of street art, making such good use of a blocked-up window in the side of a seventeenth-century building at the bottom of Christmas Steps. Well above head height, blink and you'll miss it.

Beau Stanton is a Brooklyn-based street artist best known for large murals across the States and internationally. He's not adverse to working on a small scale and in 2017 a show at That Art Gallery nearby in Upper Maudlin Street featured his paintings and stained glass. The little tile and glass mosaic was an added extra, leaving his mark on Bristol.

Years ago, I worked on the archaeological investigations behind this building making elevation drawings of the standing structures. Amongst the details I drew was another little blocked window high up on the back wall. There is something mysterious and magical about sightless windows in old buildings and Stanton's mosaic is a mysterious reprise. Like any old city location this place has been things and seen people. It's been a church, a medieval hospital, one of Bristol's oldest schools, a chip shop, a sweet shop, a clay-pipe factory, a printer's. Behind it all is a little block of flats built in the 1850s as 'model dwellings for the industrial classes.' Who looked out of those windows and watched the city change?

Ahh Toots cake shop
17 Christmas Street
(*aka* 17 Host Street)
BS1 5BT

Glazed Impression: Edward Everard Printworks
Henry Williams and WJ Neatby, 1900-1

Air pollution in Victorian industrial cities turned stone and brick buildings black. Shiny polished terracotta and glazed surfaces were one way of resisting the dirt. In the late nineteenth century, there was another factor driving an interest in tiles and that was a fascination with the Orient and the exotic – the mosques of Isfahan and Istanbul. William de Morgan's rich turquoise tiles covered the interiors of the homes of wealthy aesthetes but were far too delicate for the outsides of buildings in northern Europe. Royal Doulton and other manufacturers of architectural ceramics took up the challenge, devising brightly coloured high-fired tiles and blocks capable of withstanding both pollution and the weather.

The Everard building is faced with Doulton Carrara Ware blocks and tiles designed by WJ Neatby to reflect Everard's sense of his company's place in history. All the components are handmade and glazed, the finer detail specially modelled (one wonders how long it was before someone noticed they had spelled Gutenberg wrong). They have been carefully restored by Tiles of Stow as part of the recent conversion of the building to an hotel.

In the lane to the right, half of the side elevation survived demolition in the 1970s only through active protest. Made of Cattybrook terracotta and brick, a corner turret matches the orange glazed ones on the main front and below it a large dragon peers into the bowl at the top of the down-pipe from the roof. A central pair of dragons are now mounted in the hotel foyer. A second turret was, apparently, rescued by protesters who had climbed the scaffolding during demolition.

Clayton Hotel Bristol
(formerly Edward Everard Printworks)
35-37 Broad Street
BS1 2EQ

Carve My Name in Brick
Incised graffiti by American servicemen, 1944

In 2018, a local teenager noticed American place names – Kansas City, Missouri; Fairmont, West Virginia; Hillsboro, Texas – scratched into the brick of a garden wall in Westbury-on-Trym.

People complain about graffiti as if it is a new thing. Maybe nowadays it is just bigger. You only have to visit Pompeii or Herculaneum to see that once people could write, they began leaving their marks on the walls around them. It is about marking your presence, especially when you are a long way from home. Sometimes too, because you are fearful of the future. Cutting into stone or brick has a permanence.

In 1944 the US Army used Shirehampton golf course and other nearby open spaces for large camps to house troops arriving in Britain through Avonmouth and awaiting posting onwards in the build-up to the Normandy Landings. Soldiers who were here for longer were sometimes billeted with families around the city.

A piece by historian Oliver Davey in the local paper brought a reply from a reader. As a teenager in Westbury she had dated a GI from Hillsboro, Texas who was billeted in Falcondale Road. Hillsboro is a small place. It seemed likely the Texan brick-carver was 19-year-old Corporal Mike Willard Browning of the 300th Engineer Combat Battalion. A construction unit specialising in bridges, they were in Bristol to build the camps. They landed in Normandy a week after D-Day taking heavy casualties after a landing-craft struck a mine – but yes, later in 1945 he did come back to see her on the way home.

8-14 Henbury Road
Westbury-on-Trym
BS9 3HJ

Manilla Road, Clifton

There are few places in Bristol (or anywhere else) where the use of ceramic building materials on ordinary houses really goes into overdrive. This is one; not terracotta red, but with all the trimmings. The terrace on the south side of Manilla Road in Clifton was built in 1888. The builder must have been to see the Natural History Museum in South Kensington and, back in Bristol, rushed to one of the brickworks supplying buff terracotta mouldings and bought up their entire stock.

Jokes apart, the use of multiple columns repeated across the entire row of seven houses creates a rich wall of pattern, and must have been very bright when new. The recent cleaning of the Natural History Museum was a shock to some when it was revealed in its full glory. Here in Manilla Road very small amounts of local limestone fill the spaces between the complex ceramic mouldings. The whole thing is topped off with steep slate roofs like fancy hats and sharp dormers. They resemble more a children's picture-book house than anything you might expect to find in the street.

The popularity of buff ceramic in Bristol is partly a product of the collieries, the pale-firing fireclays being a secondary product of the mines. It was only logical to have a brickworks attached to the colliery. Sometimes these houses are the brickmakers themselves showing off. Numbers 42-44 Ridgway Road, Fishponds are opposite the site of the Fishponds and Bedminster Brick and Tile Company, and show off a range of buff clay components no doubt from their range. The red brick office opposite is now a house too.

2-14 Manilla Road
Clifton
BS8 4ED

Chimney Pots

It is easy to walk around town without ever looking up. Often, it is only when the builder removes a chimney pot and it finds its way down to the ground that we suddenly realise how big and imposing it is. The extruded and moulded pots we know best emerged in the mid nineteenth century but for a century before they were made on the potter's wheel by professional 'bigware' throwers. Two large cylinders were thrown with heavy rims – one quite simple to anchor the pot into the mortar and the other more ornate to decorate the top. Bottoms cut out; they were then luted together base-to-base to make a tube 70-120cm (2.5-4.5ft) high. You can sometimes detect a wobble where the two meet. More elaborate versions had a third section on top, crossways to keep the rain out, or side openings to reduce the tendency for smoke to blow back downwards.

Around Bristol the brickworks and redware potteries all made chimney pots. The potteries in Westbury-on-Trym made red clay ones from the 1740s, often washing the outsides with white clay slip to give them a paler, more stone-like colour. The smaller redware potteries such as the Barton Hill Pottery and the brickworks potteries in Clevedon and Weston-super-Mare continued to hand-make pots long after the market became dominated by the yellow fire-clay moulded and extruded wares of the colliery brickworks in Bedminster and to the east of the city.

For handmade pots, look at the roofs of unpretentious eighteenth and early-nineteenth-century buildings and see what treasures hide above. They vary across different areas of the city and the ages of the buildings. Clay pots were too vulgar for architects or wealthy homes. The white slipped pots have often weathered over the years and it can look at first sight as if an untidy builder has been cleaning a paintbrush on them.

Bristol Eye Hospital
Creation, Walter Ritchie, 1984-6

The large brick panels extending along the front of the Bristol Eye Hospital form what is claimed to be the largest piece of public art in the city. Creation is set out in five stages from *The Origins of the Earth* through to *Humanity*. Walter Ritchie had brick sections made in the studio and then carved them by hand as if they were stone. This is ceramic, but unusually, worked after firing rather than before. Ritchie's design uses a diagonal slant to take the eye from left to right and the occasional line jumps from one panel to another to form a link across the narrow windows and columns that separate them. The way of working varies from relief to intaglio and on the last panel, *Humanity*, unlike anywhere else, some of the brickwork has been modelled onto in clay to form the lettering. The sea creatures are sunken into the brick 'sea' whilst the birds rise out from the surface breaking the top edge of the block. The land, shown here, is more schematic, like a children's jigsaw, drawing you in to discover the insects tucked into the spaces between the animals.

The wall is designed to screen consulting rooms, placed at ground level for ease of access and with small windows for privacy. Rather than accept the resulting empty panels, architect Alan Beattie saw the spaces as an opportunity for a significant piece of public art. The panels were made in Kenilworth and brought to Bristol to be bolted onto the building. It's the way they built Asda in Bedminster, whole sections of brick arches arriving on lorries and going up in an afternoon.

Bristol Eye Hospital
Lower Maudlin Street
BS1 2LX

An Islamic Carpet

If you are in Isfahan or Istanbul you expect the mosques to be brightly decorated with coloured tiles. English parish churches are more staid places especially since the Reformation and Victorian restorers removed much of their painted decoration. A chapel paved with sixteenth-century Spanish moresque tiles is not at all what you expect.

The Lord Mayor's Chapel in College Green has long had an important place in the city and wealthy landowners and merchants made their own additions to it. Tudor courtier Sir Richard Poytnz who lived at Acton Court north of Bristol planned and began building a chantry chapel at St Mark's on College Green as a place where prayers could be said for himself and his wife. When he died in 1520 it was not quite ready and he left instructions in his will for his executors to finish the job. Someone, possibly his younger son Francis who visited Spain in 1527, clearly went to town on the flooring.

Spain was for a long time, part of the Islamic world and although by the 1500s it was controlled by Christian kings, the two cultures were able to function alongside each other. In south east Spain Muslim communities were respected for their skills in the arts, and especially for pottery and tiles. The tiles in Bristol would have been the height of fashion in Spain in the early 1500s. They would not have been easily obtained in Britain and the family would have imported them personally.

Before the Reformation, English churches would have had much more decoration and colour than we are familiar with now but nonetheless, when they were new these tiles would have been shiny, brightly coloured and totally unfamiliar to the regular congregation. The Poyntz family were showing themselves off as wealthy, connected and cosmopolitan.

St Mark's – The Lord Mayor's Chapel
College Green
BS1 5TB

Glamorous Green Roofs
Sneyd Park, c.1930

Every now and again, in the leafier parts of suburban Bristol built in the 1920s and 30s, you come across a glossy green roof on a white-painted detached house. As likely as not they were built by Stride Brothers or IW Voke. They glow in the sun and hint at holidays in the Mediterranean or adventures on the Orient Express.

This style of house was particularly popular in the seaside resorts of the English south coast with occasional examples in blue to really push the Mediterranean theme. They are the archetypal mid-century doll's houses. The tile-makers were quick to pick up on the trend. Standard red-clay pantiles were coated with semi-opaque glazes that varied subtly in colour like patinated bronze.

For the adventurous who could afford it, the 1920s heralded Modernist flat-roofed houses with steel-framed windows and smooth white walls. The architects were looking towards Europe, Le Corbusier and the Bauhaus.

The more conservative British customer wanted the new but without the scary overtones of un-English taste. Tubular steel furniture was too industrial and well, foreign. Pitched roofs, stained glass and decorative porches kept things safe whilst steel windows, white paint and brightly coloured tiles ensured that the new owners appeared suitably adventurous and cosmopolitan without taking too many risks.

Sneyd Park
BS9 1QG

Night Fighter
Bristol Blenheim. *Anon*, 2007

In the right evening light, driving into Bristol along Filton Avenue, the shadow of a wartime fighter half seen out of the corner of the eye on a high wall, momentarily suggests another time. Sometimes there are vapour trails above in the sunset. The plane is laid in grey bricks within the buff end wall of a block of flats on the corner of Fifth Avenue. An outline of a Bristol Blenheim, one of the Bristol Aeroplane Company's most iconic Second World War aircraft has been worked into the stretcher-bond of the wall.

There is a long tradition of using the different colours of standard bricks to work a decorative design into an otherwise plain wall. Tudor builders in the east of the country would save the blackened bricks from the lower courses of a firing where the direct contact with the flame over-heated them. Used artfully in a red brick wall the bricklayers made checkerboard patterns, diamonds and crosses of shiny purples and blacks. Sometimes they would add the owner's initials or a date to a gable.

Built in 2007, this work is anonymous like those of the past. A small sign describes the history of the Blenheim and records that the people of this area worked to build them and commemorates the crews that flew them. As the war becomes distant history, often described in terms of the big events, this is a gentle reminder that everyone had their part to play, that places like this were vital too. A passing local said to me that it made this small cul-de-sac somewhere special. A place to be proud of.

Blenheim Court
472-478 Filton Avenue
BS7 0LW

Heritage Passage
Anon. Bristol Development Corporation, 1994

Bristol celebrates its history in strange ways at times. Its pride in its industrial past is undoubted but if you were to look for a 65-metre long mosaic banner of highlights you might be excused for not exploring a narrow underpass down a cul-de-sac under a dual carriageway in St Phillip's.

In the Heritage Passage, two glass-tiled mosaics tell the story of Bristol manufacturing from the Middle Ages to the 1990s (well some of it anyway – sugar and tobacco are notably absent). At the centre of one side two large panels celebrate Bristol's potters. Four brightly coloured seventeenth-century Brislington Delft dishes flank a view of Pountney's pottery in Water Lane in 1820. A thrower at the potter's wheel separates them from the products and workshops of Price's stoneware bottle factory in Victoria Street in the 1880s and 1930s.

The Passage is part of the Bristol Development Corporation's remodelling of the area in the early 1990s, taking the brutal St Philip's Causeway flyover above back streets and trading estates from Old Market to Brislington. Pedestrians and cyclists were not a priority but this little tunnel was tucked in a corner linking the north and south sides of the area where the road was at ground level. Another called the Railway Passage takes the Bristol & Bath Railway Path under the road, its murals sadly submerged under graffiti.

Why and who? The Development Corporation seems unlikely to have had a particular interest in engaging with Bristol's past; they were about road building and office blocks and showed little interest in local concerns.

Heritage Passage
Days Lane, St Phillip's
BS2 0QA

Tudor Romance: Blaise Hamlet
John Nash, 1811

Elaborate brick chimney stacks looking like giant chess-pieces are a Tudor high-fashion statement most familiar from Hampton Court. Brick was a rare material in the West Country at the time, although the stacks at Thornbury Castle just north of Bristol bear the date 1514 and thus pre-date Hampton Court. Thornbury was built for the Duke of Buckingham and the bricks and bricklayers were probably brought in from London.

Blaise Hamlet in Henbury is a group of nine romantic fantasy cottages built in 1811 for the Blaise Castle estate pensioners. It was part of a project by architect John Nash and garden designer Humphry Repton to remodel the castle grounds that included a thatched dairy behind the house and folly cottages in the woods. The hamlet is arranged unevenly around a 'village green' with benches for the pensioners to sun themselves on. Each cottage is different from its neighbour and the eclectic mix of stone, brick, thatch and tile is designed to suggest evolution over time rather than new-build. The romance was more for the viewer than the resident, of course, and so the buildings had to have some element of theatre (the pensioners must have felt like goldfish at times). Hence several of the cottages sprout huge Tudor-style chimney-stacks wildly out of scale with such tiny buildings.

The bricklayers worked with standard bricks and demonstrated their skill using the corner of the bricks precisely to create huge knurled columns. They must have looked very dramatic in fresh red brick when first built. Bristol's Hampton Court? Maybe not, but Blaise Hamlet has ambition.

Blaise Hamlet (National Trust)
Hallen Road, Henbury
BS10 6QY

Bristol Byzantine Revisited
Giles Round, 2023

The final part of the refurbishment of the old Colston Hall as the Bristol Beacon has been the restoration and remodelling of the original entrance and foyer on Colston Street. Opened in 2023, the freshly cleaned façade has been revealed in all its glory, the bands of coloured brick revealed once more and light streaming from its reopened windows of an evening. When it was completed in 1873 it was a major addition to the city's public buildings. The architects Foster and Wood made extensive use of industrially produced brick and terracotta. They were supplied by the Cattybrook Brickworks set up originally to supply the construction of the Severn railway tunnel.

Cleaning has revealed the two moulded and coloured brick and terracotta friezes that form the eaves and run below the first floor windows are coloured red, green, yellow, black and white. They would have been new and novel, a bright embellishment for a place of entertainment.

Disastrous fires in 1898 and 1945, and heavy-handed repairs, had left the building in a sorry state, its windows blind and its two-storey foyer cut by concrete stairs and covered in insensitive gloss paint. Renamed the Lantern Foyer, interior designer Giles Round was commissioned to enhance it. Taking his cue from the brick friezes, he has clad the rectangular columns with brightly coloured bands of glossy ceramic tiles. At first sight the intense colours seem quite distinct from the place, but look again at the bands of bricks and moulded pieces outside and they fall into place, the geometric shapes and colours playing with the forms and colours of the originals.

The Bristol Beacon
Trenchard Street
BS1 5AR

Geoffrey the Giraffe
It's a zoo up there Chinagirl Tile for Upfest 2018

If you go down to the Sunday Market at the Tobacco Factory in Southville, you're sure of a big surprise. A giraffe pokes his head over the wall by the gate, staring calmly along the length of Raleigh Road. From inside the market he becomes a life-size figure, seemingly standing behind the railing of an old-fashion zoo enclosure. Known locally as Geoffrey, he is made of glazed earthenware, his head in 3D sections mortared together and his body of hand-made tiles. He turns a heavy freestanding block of brick and ironwork into a memorable feature of a lively space.

The area around the Tobacco Factory and North Street is the focus of the annual Upfest festival of street art and graffiti. The range of artworks is vast and changes every year as old pieces are painted over and renewed. The Tobacco Factory itself is the canvas for the largest feature paintings. Artists like Austrian Chinagirl Tile, working with more permanent materials are a relatively new part of the event. Could guerilla-tiling be a thing?

Tobacco Factory
Raleigh Road
Southville
BS3 1TF

Fragments of a Lost Museum
Doulton & Co. balustrade, c.1900

In 1769, entrepreneur Eleanor Coade acquired a small brickworks that was attempting making artificial stone. Why not, she reasoned, go a stage further and make architectural parts that could replace costly carved stone? Using light-coloured high-firing clays, Coade devised a smooth, tough weatherproof material close to limestone in colour using moulds to allow multiples to be made rapidly. Marketed as a sophisticated new product Coade Stone was very successful through into the 1840s. It can be found as components of major buildings and as sculptures such as the lion and unicorn at the gates of Kensington Palace.

Brown's brasserie on Queen's Road started life in 1872 as the City Museum. Modelled on a Venetian Palazzo it was a striking building and welcomed visitors to see its famous collection of fossils amongst its other attractions. Left partly roofless after the Blitz in 1940, the fossils decimated, the surviving collections were moved to the Art Gallery. Patched up with concrete, shorn of much of its architectural detail, the building itself spent many years as a student canteen.

The fragile-looking balustrade in front has somehow survived the building's adventures. It loses columns from time to time – a sort of ongoing dental problem. Although made by Doulton of Lambeth much later, they are Coade Stone in all but name, performing the same role as a ceramic replacement for costly carved stone details. Some of the repairs involve refitting the originals but it has 'false teeth' made of stone and cement. Eleanor Coade would have sighed at the thought of a stonemason being asked to make Bath Stone replacements for them.

Brown's
38 Queen's Road
BS8 1RE

TH ·ARE · DEPO

EDWARD
DIED · NOVR.
OF ·MARTHA
DIED ·MAY

In Memoriam

In the corner of the south transept of Bristol Cathedral a few steps lead down to the cloisters. Beside the steps a tiled panel commemorates Edward Bird and his wife Martha. The gothic lettering is hand-drawn in red and black onglaze enamel on white earthenware tiles. The looseness of the drawing and the slightly awkward fit with the tiles seem to be deliberate rather than clumsy. They are not a commercial product and were probably done through a local enamelling workshop.

Edward Bird was a well-known figure in Bristol, a successful painter who was elected to the Royal Academy in 1816. His friendship with other local artists led to the emergence of the Bristol School of Artists who met regularly to paint and draw together in the first decades of the nineteenth century. Many went on to great success themselves including Samuel Jackson and Francis Danby.

Despite the quirky choice of utilitarian white tiles, this is fashionable medievalist taste at a time when Pugin was working with Mintons to produce reproductions of medieval church tiles and reviving period lettering styles. The Birds' daughter Harriett emigrated to the United States in 1849 describing herself as an artist. She and her daughter were later active as artists in San Francisco. It would be nice to think Harriett painted the tiles herself as a tribute to her parents.

Bristol Cathedral
College Green
BS1 5TJ

Some Bristol Potteries

There have been dozens of pottery and brick and tile businesses in Bristol across the years. Very few have left any trace but we know there were several key locations.

Wootton Road, Brislington

The ceramics industry in Bristol began in the 1650s at Brislington, when Quaker delftware potters from London moved to England's second city. The pottery was at the western end of Wootton Road in St Anne's. Persecuted as Quakers, being outside the city they were beyond the reach of civic authorities and near enough to the port.

Temple Back, Victoria Street and Redcliff

By 1700 the successors of the Brislington potters had moved into the city and congregated around Redcliff and Temple: Thomas Street, Victoria Street and Temple Back in particular. The area remained the hub of the pottery industry until the beginning of the twentieth century and together with the glass factories would have looked a bit like part of Stoke-on-Trent. The industry shrank steadily in the nineteenth century and by 1900 the main survivors were Pountney's making china and sanitary wares and Price, Powell & Co. making stoneware bottles. Price's works was destroyed in the Blitz in 1940.

Lodge Causeway, Fishponds

In 1905 Pountney's moved to a large modern factory at the western end of Lodge Causeway in Fishponds. According to locals the hollow spaces in the railway bridge there are filled with the pottery's cast off plaster moulds and kiln waste. They closed in 1969. The pottery has gone, but not the pots. The best Pountney's plates have brightly painted floral patterns based on early Bristol delftwares and were made in the 1950s.

Brick seconds used to make the entrance road to the Cattybrook brickworks
Ibstock – Cattybrook Over Lane Almondsbury BS32 4BX

The Smaller Potteries

Another eighteenth-century delftware pottery was located where the primary school stands in Limekiln Lane, at the bottom of Brandon Hill. Potsherds sometimes appear between the cobbles there.

The Sugar House Pottery in Westbury-on-Trym supplied the sugar industry between 1750-1860 and competed with potteries like the Barton Hill Pottery in Queen Anne Street, Barton Hill; and the pottery in Boot Lane, Bedminster making plant pots and rhubarb forcers. In the second half of the twentieth century came the studio potters. In the 60s and 70s The China Factory (Eric Wayman) and the Rainbow Pottery (David Thornley) in Clifton were well known. The China Factory's large slip-cast Morris Minors were very popular, even gracing the collection of Elton John. The workshop with its tiled sign still exists in Vyvyan Road.

Cattybrook Brickworks, Almondsbury

The Cattybrook brickworks at Almondsbury is not quite in Bristol, but of all the brickworks that built the city it's the only one still operating today. It's now part of the Ibstock group but its history goes back to 1865. Established by an enterprising railway engineer, it supplied more than 70 million bricks for the Severn Tunnel. In fact their bricks are all around us, from the Tobacco Factories in East Street and North Street to The Granary on Welsh Back. If it is red brick around Bristol, then it's likely to be Cattybrook. Today's Cattybrook range includes Bristol Red, Bristol Gold, Cheddar Brown and the lovely sounding Bristol Orange Blend.

Contemporary Potters

There are many studio potters in Bristol today. Some exhibit nationally and internationally and others show their work through local galleries, markets and the many arts trails. The Peoples' Republic of Stokes Croft in Jamaica Street are more factory than studio; they are best known for their political bone-china mugs as well as a range of tablewares decorated with a collage of recycled mid-century transfers. An Edwardian fish shop, the PRSC's showroom has a splendid tiled front featuring a battered sea bass.

Brickworks

Bristol is a stone city and brickmaking did not take off until the middle of the nineteenth century. Most brickworks in the city were in the coal-mining areas of south and east Bristol often run by the collieries themselves. Many were large – most of the trading estate at Ashton Vale opposite the Bristol City stadium was a brickworks and the steep slope up to the A370 is the face of the clay pit. Other large brickworks were in Crew's Hole, Hanham, Fishponds and Warmley. The offices of the Fishponds and Bedminster Brick and Tile Co survive as a house in Ridgeway Road, Fishponds.

Postscript

Inevitably the city moves on and things disappear. Sometimes behind paint or boarded over shopfronts, tiles or decorative terracotta peak out. Quite a number of recent artworks around the city have been obliterated by tagging or painted over. The Fishponds brick fish and the Rodney Harris relief-brick panels of animals on the river side of the RSPCA dogs' home are sad examples. The huge glass-mosaic railway station mural in the underpass that takes the Bristol to Bath cycle path under the Relief Road at Lawrence Hill dates from the early 1990s but has been completely obliterated with paint and used as a graffiti wall. The two Relief Road underpass mosaics are arguably Bristol's largest public artworks. In a city like Manchester works like these would be and are celebrated, but it seems not here.

Sometimes the story is different. There follows two examples of ceramic works that have moved on but remain loved. The camel has gone into a loved retirement and the Froomsgate Map (overleaf) is on display in Bath.

The Cotham Hill Camel was for many years a feature of Falafel King's window. Their move to Broad Quay signalled his retirement. He was made by Margaret Power, ceramics graduate of Bath Academy of Art in Corsham and a respected teacher of people with learning difficulties. Margaret was a regular customer at the café. One day she arrived with a box in her arms: a present. Not a bad thank you for excellent falafels.

Lost Ceramics – the one that got away
Froomsgate Map, Philippa Threlfall and Kennedy Collings, 1971

In the 1960s Bristol planners envisaged a 'city of the future' with pedestrians and shops raised to a second-floor deck above the world of traffic, delivery bays and parking. Over-ambitious and destructive, the scheme fell out of favour and the disconnected pieces became empty or inaccessible. The only completed section spanned Nelson Street, Rupert Street and Lewin's Mead. By 2014 it was being progressively dismantled. Above Rupert Street, in one corner of an empty piazza was the glazed entrance to Froomsgate House, a bland 15-storey office block. The foyer wall was covered by a large map of the area in handmade ceramic tiles by Philippa Threlfall and Kennedy Collings. It reproduced part of Jacob Millerd's wonderful map of Bristol published in 1673. Millerd's woodcut shows every building and back garden in detail. He even did updates, tiny buildings cut and dropped into the printing blocks. The mural mimics that approach, using multiple stamped elements to build up streets, lit up with bright glazes and pools of melted glass.

Was there a conscious irony in the busy colourful intricacy of the map and its representation of what the utopian, out of scale development had erased forever, placed against the plate glass, pink facing-brick and concrete? Walking through the empty decks, finding it was like finding a single flower growing out of the pavement.

In 2014, shorn of the connecting bridges, Froomsgate House was reinvented as student accommodation. Rescued by Philippa Threlfall's son Daniel Collings, the mural is now housed in his Bath showroom.

Once Rupert Street, now
Black Dog Tiles
Meadows Lane, Bath
BA1 6FB

More to Discover

Booth, M 2020. *111 Places in Bristol That You Shouldn't Miss*. Cologne: Emons Verlag

Gomme, A, Jenner, M and Little, B 1979. *Bristol: An Architectural History*. London: Lund Humphries

Henrywood, RK 1992. *Bristol Potteries, 1775-1906*. Bristol: Redcliffe Press

Merritt, D and Greenacre, F, 2011. *Public Sculpture in Bristol*. Liverpool: Liverpool University Press

Morris, S and Mowl, T, 2001. *a Bristol Eye – the city seen from new perspectives*. Bristol: Redcliffe Press

Reg Jackson, Bristol Potters and Potteries: https://www.bristolpottersandpotteries.org.uk/

Valda Jackson and Rodney Harris: https://www.jacksonandharris.co.uk/

Jo Amey *aka* The Tiles Lady: https://tileslady.wordpress.com/

Black Dog Tiles: https://www.blackdogtiles.com/

Chinagirl Tile: https://www.chinagirltile.com/

Philippa Threlfall: https://www.philippathrelfall.com/

Acknowledgements

Heather Kent for everything; Stephen Morris for his patient guidance; Matt Benton, Mark Samsworth and Oli Timmins for tipping me off about places to go; Jo Amey (The Tiles Lady), Oli Budd, Julian Brenard, Daniel Collins (Black Dog Tiles). Oliver Davey (Yanks in Bristol), Falafel King, Peter Foreman, Rodney Harris, Hilary Irvine, Bee Lang for proof reading, Theresa Searle, Philippa Threlfall and Jo Young for their help and Stephen Morris and Heather Kent for additional photography.

About the author

Oliver Kent studied Ceramics at UWE and Ceramic History at Staffordshire University. He has taught ceramics at all levels for many years principally at Bristol School of Art. With a background in archaeology as well as making he is interested in every aspect of his subject from a brick building to a sherd of broken pottery, how they are made and what it might be used for. He has written widely about historical ceramics and design. As a maker his work combines both history and art and is in several public collections in the UK.